INVENTOR'S INCUBATOR:

MY FIRST PATENT

REGGIE D HARMON

QuestWorld.TV, Inc.

This book is dedicated to the creative people of the world that go unnoticed and forgotten.

"The Day Science begins to study non-physical phenomena, it will make more progress in one decade than in all the previous centuries of existence."

NIKOLA TESLA

CONTENTS

PREFACE

I've had a great deal of fun and have gained new inspiration in writing this book. I hope that you will find it exciting to read and it will help you to see new possibilities in your life.

Maybe this will be the catalyst that lifts you above the challenges of everyday life and leads you on the course you've always dreamed of. I wish that for you.

INVENTOR'S INCUBATOR: MY FIRST PATENT
By Reggie D Harmon

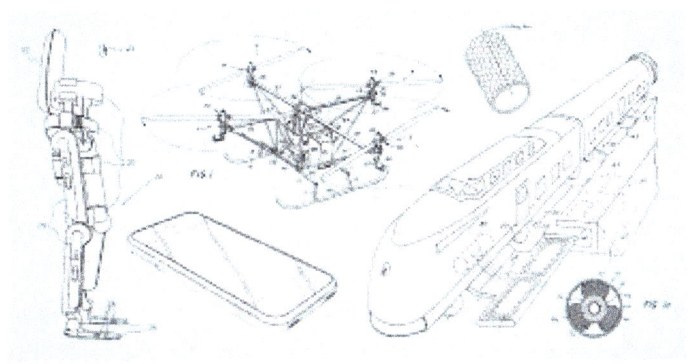

CHAPTER ONE: A STARTING PLACE

Wouldn't it be amazing if all the people of the world were free to create and invent amazing things? Of course, most of us have life issues keeping us from true freedom of creation. All the stuff we cram into our daily lives and then repeat the next day holds us back. I'm going to invite you on one of my journeys to explore the world of invention. I like to delve into things as a complete novice and try my best to succeed without professional help or guidance, which I do not recommend other people to do, so there will be a lot of mistakes made on this adventure, guaranteed and this is certainly not a how-to book. Ever heard of "If all else fails, read the directions"? The path this takes us on will not be a direct path, you will not get from point a to point b in the most efficient manner, but hopefully you will enjoy the adventure and learn from it.

For instance, I'll give you an example of my annoying daily life stuff, which was my truck not wanting to start. The starter would barely turn over and the truck would start by the skin of its teeth. I ran different diagnostics to determine if it was the battery, or the alternator, or the starter, or the wiring. While the engine was off, I measured the battery voltage with a meter, then with the car running, I could see the voltage was higher, 14 volts. Probably, not the alternator. Then I used a jump-starting charger to try to start the truck and to my surprise it wouldn't turn over at all. It made the situation worse. Bad words followed! Now, I had two possibilities that were the most probable culprit, the wiring or the starter. While I was measuring the battery voltage, I had a friend start the truck and the voltage dropped to 11 volts, so pretty sure the wiring was good. My likely suspect now was the starter and after replacing it, I have a truck that starts again and I have one less life issue. Annoying story, can you feel an empathy with all the stuff that happens to you? All the mundane things you need to get done everyday just to be able to go thru more of them the next day.

The point of all that is, I like to use trial and error to solve most things. This means that there is going to be failures along the way in most of the things I undertake. Knowing that in advance, makes a person less emotional about the failures and more focused on reaching the final outcome, which we hope is success. And, what does success look like for you? I mean, success beyond the daily grind. You have to decide. After all, Thomas Edison's teachers said he was "too stupid to learn anything." He was fired from his first two jobs for being "non-productive." As an inventor, Edison made 1,000 unsuccessful attempts at inventing his light bulb. What a failure, right? Most people would have given up after a few dozen tries, maybe less. Glad he kept trying even when his teachers told him he would never amount to anything. And, even at that, he didn't really invent the incandescent light bulb, he improved on previous inventions of it.

The incandescent bulb existed, but they were expensive and burned out quickly. Edison's invention, his light bulb, solved both of the issues.

This book's objective is not to give you a precise tutorial in getting a patent. You will want to seek outside help for that. You will find tons of material on the internet and from other sources that can help you do that. But be wary of many of them. There are also tons of scams out there. We will end up discussing how I applied for a Provisional Patent on my own and received a Provisional Patent. This is actually not that difficult of a task and you should get an overview here that can be helpful in choosing your path to a patent. And I'll talk about my applying for a Non-Provisional Patent before we part. But, will there be defects in my patent applications? Probably, but remember I take the path of most resistance quite often, but I enjoy the learning curve associated with doing so.

What this book should do, is stir your creative side. You should already be thinking "I can create something unique, I have some really great ideas." And even if you don't have a precise idea now, I'm going to show you how to get inspired and find an idea. We will look at inventions from all over the world and it will be fun. From these inventions, you may find you get inspired and creative.

I didn't really choose my invention, it chose me. I really like music and I like mechanical things. Vending machines, robots, slot machines and jukeboxes all fascinate me. A jukebox doesn't make much sense today. I mean, the mechanical ones that take a CD or 45 rpm record from a magazine and plop it down on a spinning platter. You can have more music on Spotify or a thumb drive, right? But something about seeing this mechanical arm picking up the CD and putting it on the platter is cool to me. And the lights and great sound quality doesn't hurt either. And believe it or not, they still make brand new jukeboxes today.

So, I ordered a new jukebox from Rock-ola. They make them by hand and it took several months to receive mine.

The delivery truck dropped it off in my driveway. It took four of us to take the 360 pound thing up a flight of stairs and then down a flight of stairs safely. Plugged it in and put a CD in and it was like stepping back in time. Great sound, fun watching it place the CDs on the platter, but one thing didn't work for me. The device that listed the album tracks, or song selections. This is a motorized, mechanical thing that flips plastic pages in which you insert a paper list of each song selection. The mechanics are kind of cool and I'm sure the nostalgic purists love the mechanics of it, but it's limited to how many tracks you can list and if you decide to fill the jukebox with 100 CDs, then it's even more limited and the track listings become even less readable.

This was where I started thinking, I can write some code and I have some electrical and electronic knowledge, maybe I can create a solution. And I did. Of course, I didn't want to cut into wiring harnesses or alter a brand new jukebox, so my solution had to avoid that. It had to be a module that replaced the original factory module and looked factory made.

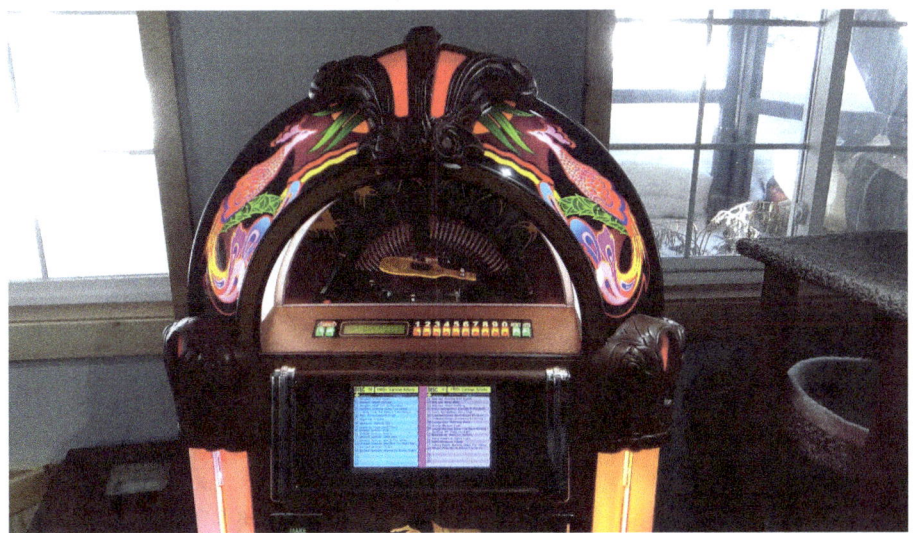

I'm not going to give you specifics or drawings and go into the whole nuts and bolts of my invention here for a number of reasons. One of them is that my Non-Provisional Patent application is in progress. The typical time to perfect a Non-Provisional Patent is 22 months, almost two years, so it will be a while before I want to discuss specifics. I do have a Provisional Patent, but as we will see later, I do not want to depend solely on it.

I assembled what I thought should be all of the components and created a prototype. I put it into the jukebox and it failed to work properly (of course!). What in my world works perfect the first time? Or even the fifth time. Very little! Remember, trial and error. Not too many inventors get their invention to work the first time, so why should I? The jukebox's processors expected certain things to happen and they weren't. I had no idea what the jukebox's expectations were. No documentation or information was easily available to me. I had to try different things to see what worked. This caused me to go back to the literal drawing board. But I wasn't starting from scratch. I had gained more information from trail and error and just needed to add some code and electronics. Still a couple more failures including CPU overheating and then finally a success. I had a prototype that worked!

Now, I went to work on a business plan and bought several of each component and geared up to create a production line for these modules. What would be the market for them? I knew, that these jukeboxes have been sold since the 1980s, so my guess is there's thousands of potential customers. Awesome, I would sell them on eBay. I calculated a price point and profit margin. Then I decided how many of them I could build per day, per week and if this could be profitable enough to be a sole vocation. It might be. But wait, if I was successful, people might copy my designs and go into competition with me. That would change all of my calculations. How could I solve that issue? I decided I needed to get a patent.

I've heard of people getting patents and seen commercials that will help an inventor, but I wanted to try my hand at it. I also didn't trust many of the advertisements I've seen that "help" inventors get patents, or purport to help inventors. Still unsure of the true potential for sales of my invention, I needed more information and options.

The first thing I did of course, was turn to the internet and do a search. All kinds of stuff came up including, on line companies that would "help" me file my patent. Some of the sites I found were also scams or insincere, how could I figure out what to do? One of the links was The United States Patent and Trademark Office (uspto.gov). I had to trust this site as it is the official site of the government entity that would grant me a patent. All of the information on this site is free, except for filing fees of course. They even have a learning and resources section on their web site. From this site, I learned that there were three types of patents; Utility, Design, and Plant. My invention fell under the Utility patent heading which may be granted to "anyone who invents or discovers any new and useful process, machine, article of manufacture, or composition of matter, or any new and useful improvement thereof;"

Since June 8, 1995, patent applicants can file either Provisional or Non-Provisional applications for utility patents. The Provisional Patent is obtained much quicker and the process is extremely streamlined.   It allows an inventor a level of protection, while giving the inventor time to decide if they will want to file a Non-Provisional Patent. The twelve month period that a Provisional Patent is good for, also allows an inventor a chance to get investors interested and creates a level of protection for their idea.   This also allows the inventor to engage in other activities to get their invention off the ground without endangering their claim (if the application is perfected properly).

When I started going thru the Provisional Application, I found that I really didn't understand my own invention, even though I was looking at a prototype! I really needed a simple place to start. There are some basic questions that allowed me to sum up my invention. I'll name a few of them to help you. What is the purpose of my invention? Describe the problem(s) that your invention solves? How is this invention an improvement over existing technology? What individuals or businesses will benefit from or use the invention? What are the specific benefits of the invention to its users? I answered these questions and wrote a brief description of my invention which helped a great deal. If you don't have an idea for a patent, fear not, think about something you're interested in and you will have almost limitless opportunities to create something incredible.

This is where the true exploration starts! I went to the Patent Public Search page. This is a free patent search tool provided by the USPTO (United States Patent and Trade Office). You can reach it thru their main page, or google it, bing it, or https://ppubs.uspto.gov/pubwebapp/

I ended up creating an account at my.uspto.gov, which again is free. You can register as an inventor and later this account will be useful if you decide to submit a patent application directly. Creating an account is easy and it is the first step in filing patent applications. Doing it was a concrete action toward my goal. Each small step helps inspire me. Taking small steps may inspire you. Creating an account has no obligation or down side that I know of, and just that small step could inspire you like it did me.

But back to the Patent Public Search. If you don't have the mega million dollar idea yet, this is a great place to get one. If you already have the idea and have answered the same questions I did, then this is a great place to see if someone else thought of it before you. Searching patents is free and can save you a great deal of money and avoid wasted time working on something that has already been patented. Some of the patent filing services will file a Provisional Patent for you without a search. You will get a patent, but it might be useless. Even more troubling it might infringe on a previous patent. If you decide later to file a Non-Provisional Patent application that cites benefit from the Provisional one you filed earlier, and if the Provisional one doesn't have a real basis, you will need to abandon it and start fresh, losing your original application date. This can be a serious issue if you have a really marketable idea. Conducting a patent search also gave me a great deal of clarity about my invention. What I did, was create search terms and then search thru patents that may have the same technology or basis that my invention does.

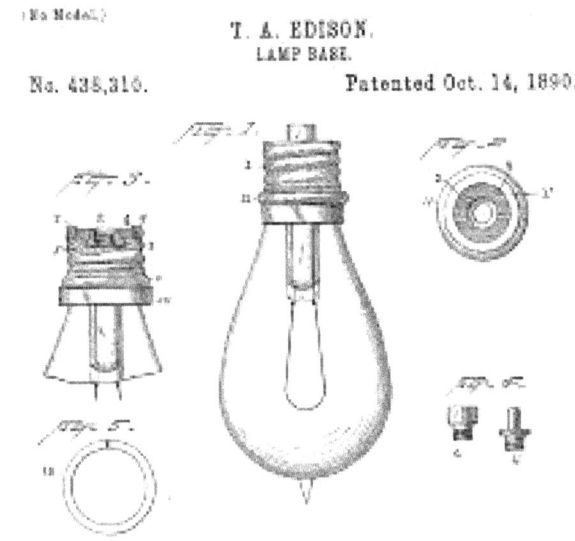

Think about Thomas Edison's bulb. It really wasn't completely unique because several patents had been granted for incandescent lights before his. And didn't all of them have electricity and wiring in common? There were many other similarities between his invention and previous ones. What truly made his invention patentable was the improvement of creating a vacuum in the glass structure, which created a huge improvement in the supplied light. If I were Thomas Edison and conducted a search I might type in electric AND bulb AND light, OR electric AND light and get all sorts of inventions, not just incandescent bulbs like his. This is where filtering and constructing the best search criteria comes into play. In looking thru all the inventions that appeared from the search, though, I'd find some pretty interesting inventions that would inspire me and see the specific format for a patent. I'd also see the claims section that describes how each invention improves on the previous ones. This gives me more ideas and honestly looking at some of the past inventions is really fun. It's like stepping into the shoes of another inventor for a few minutes. The closest thing I have to time travel right now!

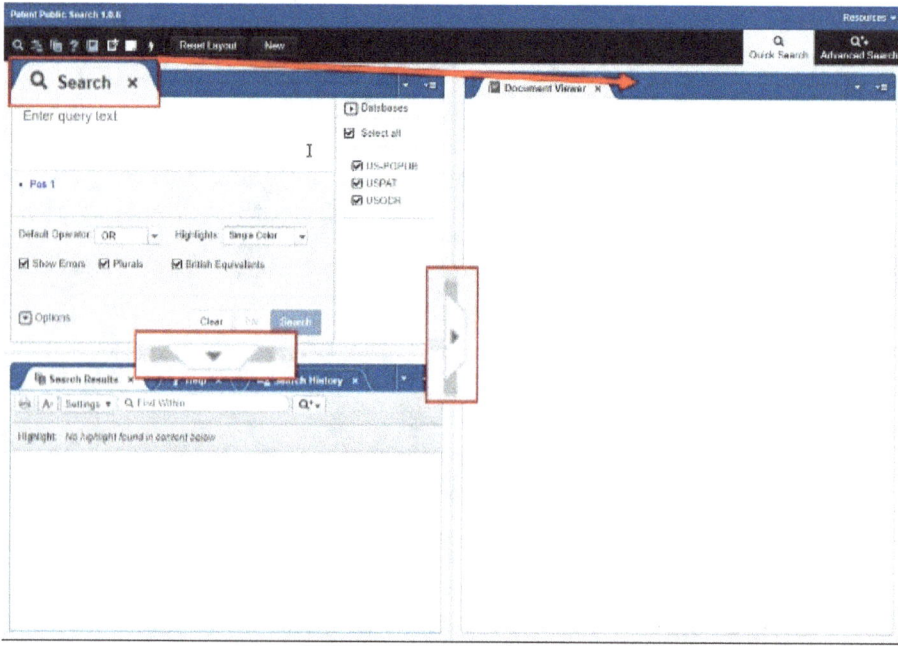

In my case, my patent search had to include language and terms that aren't used today. It will probably be the same if you do a search. You have to think what other names did they call your device or invention in previous years as well as various names used for it today. The patent search tool requires a special search string to be entered to eliminate as many patents that do not apply. The search tool has AND and OR operators (and some others), so you can construct a search criteria that filters most patents out and gets more of the ones you want to see. You don't need to be afraid at trying out some searches and seeing what you get. Then, change up the search string and run your search again. Putting terms and one of the operators in parenthesis means the search engine will consider that equation first. My search string looked like this: "automatic AND phonograph ((title AND selection)) OR automatic AND phonograph AND menu OR jukebox AND ((title AND selection)) OR jukebox AND menu". Of course, you leave off the quotes and period at the end. I put this into the upper search window and clicked search. This brought up patents for all kinds of storage devices, music players and more in the search results window below. There were over 5,000 patents to sort through. In the search results window, there are columns that you can make marks in. You are basically putting check marks beside each of them and deciding their level of relevance. Try it out, go to the search tool and if you want, use my string and substitute some terms that are associated with your invention. See what happens. You could stumble upon some interesting stuff you never thought possible.

For me, I went thru and checked the select box next to the list of patents that seemed very similar and also checked the boxes based on relevancy; with a 1 if quite similar, and those using less similar technology with a 2,3,4, or 5 depending on how strong I felt they were relevant. You can experiment and create check boxes in the search tool to help your search. In order to get thru all of the displayed results fairly quickly, I just looked at the title of each patent that came up in the search results window.

I found using the search tool was somewhat cumbersome and confusing but I pushed thru anyway. You may find it easier for you or discover a better way of using it. I found I could use the arrows between the search results and the search area to pull down or up and make the area I was working with at the moment bigger or smaller. The arrow in the middle portion between the document viewer and the two search boxes could be used to widen the search results area and show more information. In fact, it can be slid all the way right so that you can see the patent titles, thus making decisions about relevancy quicker. It took some experimentation and frustration for me to find the best way to use the tool. If you find it frustrating at first, just keep running searches and experimenting with it. It will get easier with experience.

Once I had all my selections made, I made a subset out of them by hovering over the select tab and right click which brought up some options. The one I was most interested in was "Create L# from selected documents", which I highlighted and did. Then I went into the search results and could see way fewer patents on that search line. Nice! Then I printed the search results for a couple of reasons. To use of course, but I also wanted to document my search efforts. It's important to me to have dated notes, drawings, and document work product in case I needed to defend my patent in court. I know, having dated material when it was created could be helpful in defending an invention. If you have a really valuable idea, it may come to that someday for you as well, so documenting everything you do, probably has no down side.

Now that I had the search results culled to a point that they were manageable, the next thing I did, was to go thru each one and read the actual patent. This was a lot of fun. It's like stepping into the shoes of each of the inventors and getting a real experience of their thought process. This really jump started my creative process and if you're looking for a place to begin your journey of development, it might be the same for you.

So now, by widening the document viewer, in text view I could see the document ID, date published, inventor information and more. In scrolling down, a field exists that is very important. This field is the "CPC CURRENT". The CPCI and CPCA codes can be used to search patents worldwide. This will be needed if I want to file a patent in another country. It's also a great way to narrow a search on the USPTO site. I wrote them down and took screen shots. The thing I wanted to do, was to see patents that have the same CPCI. That would be my main search criteria for a later world search. It is also great for doing a follow-up search on the USPTO site. My search criteria would then just be the CPCI code(s). That is way more efficient and easier than searching using search terms, but you have to start somewhere to find the CPCI!

The next sections in the displayed search results display the abstract, which is a summary of the invention, and then a background which might contain some of the claims of the invention. The claims in a patent are very important. How the claims are worded or crafted determine the strength of protection for the patent. This is one reason why people pay patent attorneys huge sums of money, to craft the perfect claims of an invention. There are companies that exist whose sole purpose is to find ways to craft a patent around another patent's claims. Their objective is to give their client a similar patent without infringing on a previous patent and to not pay the previous inventor anything for their idea. If you have the billion-dollar idea, you will want to seek out an expert to try to protect you from these attempts to benefit from your idea without compensation. Hopefully, you will find enough information in this book and thru other sources to weed out the patent scam artists from the true experts and be able to select a truly legitimate firm or individual to help you. And, hopefully by reading this book, you may be able to better summarize your invention and determine if it has true economic value before spending large sums of money.

In the document viewer, upper left, switching to image view, I looked at the actual patent filing. This allows me to see the construction of a patent, which was great for me because I wanted to file my own patent and didn't have any idea where to start. It also has the "art" or drawings, which are normally required as part of the application. The patents that had similarities to my patent filing I printed, either on paper or to pdf. Some of the ones I printed to paper were useful to hold in my hand and read thru. They actually made interesting reading!

The USPTO search is but one search tool. A patent from the USPTO is for the US only, most countries have their own patent system. There are many more search tools, including Google at patents.google.com. You may even want to start your search with Google. It is definitely easier to use than the US Patent search tool, but I found the US Patent search tool gave me better information after running a few searches and fine tuning my methods. In addition to actual search tools, you can search markets for inventions. Anything that is already in the public purview is not patentable, so the wider a search, the better the chance a person has of securing a patent and creating claims that are unique. It is also important not to infringe on a previous patent. The search for similar inventions can include on line sellers like eBay and Amazon. Seeing what is already in the market place is both important from a marketing standpoint and patent application position. Regular search tools can also be used to see what results appear. This can also give a person additional search filters and criteria for using other search methods. A thorough search for filing a Provisional Patent Application is not required however, because they are not formally reviewed, but for a Non-Provisional Patent, a wide search will be conducted by the patent examiner. I thought it was wise to conduct my own search before making an application that would be rejected. This could save me hundreds of dollars in application fees as well as a great deal of time wasted by submitting an application that was sure to be rejected later. Many of the on-line scam sites will promise a Provisional Patent and they certainly can deliver on that promise because Provisional Patents are not formally reviewed by the Patent Office. However, later, using that Provisional Patent application filing date for benefit on a Non-Provisional Application may not work, because the patent most likely, would be rejected due to prior claims or other filing defects.

Another helpful and free search tool is Espacenet at worldwide.espacenet.com. This tool searches inside and outside the US and is easier to use in my opinion than the USPTO's.

When I did the search on the USPTO's site, I wrote down all the CPCI and CPCA codes and hoped that the most pertinent patents would have a few of the same codes. Then, I went back to those codes and input the similar ones as a basis for the Espacenet search. This filtered the search and I got a much more selective set of patents to look thru. Once again, digging into some of these patents, provided me pure inspiration. I could see how the inventor, or more likely, the inventor's legal counsel constructed the patents. More enjoyable for me, is looking thru the art (drawings) and claims and comparing them with my invention. This allows me to refine some of my claims and consider the important aspects of my invention. Could it lead to a massive improvement of my art? Of course, it can!

Once I decided that my invention was unique. I was ready to file a Provisional Patent Application. Although there was similar "art" in previous patents, my claims were different and I found my design made improvements over prior art. A Provisional Patent is a less expensive and watered-down version of the Non-Provisional Patent or Regular Patent. A Provisional Patent is good for twelve months, creates a submittal date, and allows an inventor some added options for revenue raising, market analysis etc. before filing a Regular Patent. It is not formally reviewed, so most Provisional Patents are accepted, meaning you can gain a Provisional Patent that may not be worth anything. But the cool thing is you will have a patent! If that's all you're going for, is to be able to say you have a patent, then filing a Provisional Patent yourself is the inexpensive way to go. You can tell all your friends you have a real patent and are a bona fide inventor. If you really have that great idea and want to use the Provisional Patent priority date, then you should get some expert help. The reason being is that when filing a Regular Patent and referencing a Provisional Patent's priority date, no new claims can be made. The claims made in the Provisional Patent will be the claims made in the Non-Provisional or Regular Patent Application. Remember, how I said the claims are extremely important? I'm sure my claims are full of holes because I haven't written hundreds or thousands of them and I'm not a patent attorney. It's advisable to seek out an expert that knows all the pitfalls of filing applications and crafting claims that stand up in court proceedings.

CHAPTER TWO: THE NUTS AND BOLTS

I will tell you my process, so you can get a general idea of the process. The first thing I did, was to go to uspto.gov and click on forms. Then under "Customer Number", I choose, "Request For Customer Number" and filled out the form and sent it in. In about a week, I received a customer number. You can file without it, but it's nice going forward to have established an account. You can also have them email your customer number which should be faster.

The next thing was to see if I could qualify for Micro Entity status. The fees are greatly reduced for Micro Entities. Looking for form SB/15A and filing it with my application saved me 75% on filing fees.

Then I downloaded the rest of the forms; Utility Patent Application Transmittal Form (AIA/15), Provisional Application for Patent Cover Sheet (Form PTO/SB/16), Application Data Sheet (PTO/AIA/14). I filled each of these out and printed them.

Next, I created a Specification, claims and abstract file as shown below:

The Specification is a written description of the invention and how it is made and used. The first page contains the title of the invention, which should be as concise and specific as possible and limited to 500 characters or less. Next, is the "Background of The Invention", this may contain statements of problems or deficiencies of prior inventions. The third element in the Specification is a "Brief Summary of The Invention" this presents the main idea of the invention and how it solves the problems listed in the previous section. Drawings of the invention are attached to the end of the application packet. In this next section "Brief Description of Drawings" all the figures have numbers to depict each of the elements contained in the drawings and each is described here. Then, the "Detailed Description of the Invention" requires a description that would allow any person of ordinary skill in the field of the invention to be able to make and use the invention. Each element in the drawings should be mentioned in the description.

The next section is the "Claims" section. Remember how important this is? I want to keep reinforcing it! It is so important that it gets it's own page. The claims section defines the scope of protection provided by the patent and should be carefully crafted by someone well versed in doing so. Here, if anyone is in their right mind and have an invention of importance, they should consult an expert. But since I'm neither of a right mind and don't have a hugely valuable invention, I'm proceeding on my own just to experience the process and have some fun at the same time. For my claims, I listed what was unique to my invention.

On its own page next, was my "Abstract of Disclosure", this is a narrative of the invention and what is new about the invention. It is generally one paragraph and limited to 150 words.

Lastly are the drawings. Can't draw? I didn't let that stop me. With an 8 ½ x 11 inch diameter paper and templates from an office store, I made drawings. The drawings have specific format requirements and even a type of paper and type of ink requirement, so consult an expert or at least read thru the USPTO guidelines carefully. On my drawings, I had to have a top and left side margin of at least one inch, a right margin of at least 5/8 inch, and a bottom margin of at least 3/8 inch. I just made all the margins 1 inch to make it simple for me. I numbered each drawing starting with "1" at the top below the 1-inch margin. Each figure was then numbered FIG 1, FIG 2, in $1/8^{th}$ inch letters and each element of the figure was then numbered. Looking at other drawings from my patent search gave me examples, but there are more specific rules in the "Nonprovisional (Utility) Patent Application Filing Guide" on the USPTO web site. Although, Provisional Patent applications are not scrutinized like Non-Provisional ones, I intended on filing a Non-Provisional patent at a later date and I intended to claim benefit from the Provisional one, so I wanted to try to meet the Non-Provisional patent guidelines with the intent of avoiding an issue later.

For a Provisional Patent Application, the filing fee is the same if you web file or mail it in. Web filing involved getting a notarized form and I didn't want to bother with it, so I chose just to mail it in. In order to get a receipt, I enclosed a 4 by 6 inch post card with postage on it. I just stuck a forever stamp on it. On the post card, I put the filing date, my name (as inventor), and the title of the invention. Then below that on the same side, I listed everything contained in the filing packet and number of pages like this:

Check # 1234 in the amount of $ 75.00

Provisional Application For Patent Cover Sheet - 2 pages

Application Data Sheet – 9 pages
Specification – 10 pages
Claims – 2 pages
Abstract – 1 page
Drawings – 4 pages

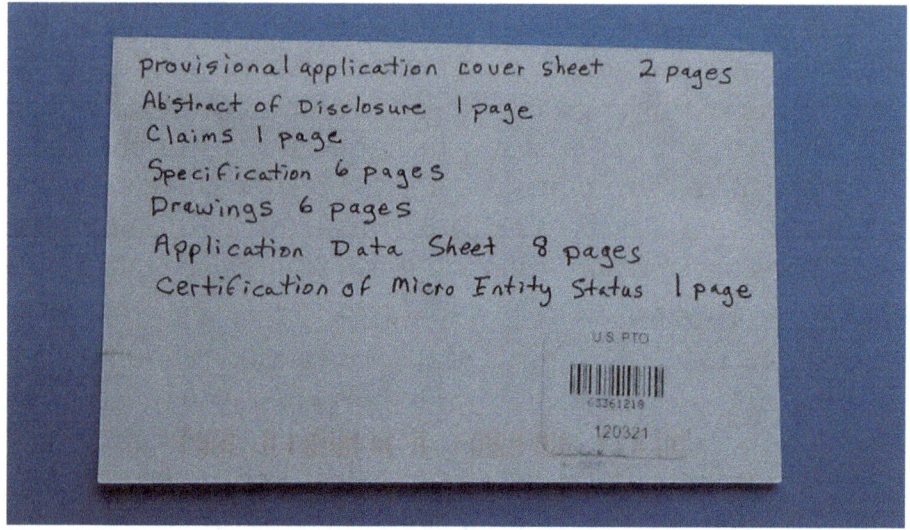

On the other side I put my return address in the upper left of the card as the sender and then my return address in the mail to area as well.

In filing by mail, I needed a Priority Express Mail envelope which I obtained from the Post Office. I chose the large Tyvek envelope. Also, I opened a "Click N Ship" account with the Post Office on line the day I filed and just went into that account and selected "Priority Mail Express Flat Rate" and printed a label with a shipping number. The shipping number was super important. I had to enter that label number on the "Provisional Application Patent Cover Sheet". The address to ship to was Commissioner for Patents, P.O. Box 1450, Alexandria, VA 22313-1450.

I assembled the packet in the order written on the post card, except, I put the check for my filing fees first, then the post card, then the cover sheet and rest of the items listed on the post card. I sealed it up and took it to the Post Office. Another very important thing to get is a mailing receipt stamped with the current day's date when you drop off the envelope at the Post Office, I made sure not to leave without the receipt. This shows a filing date for legal purposes. I also enabled text updates so I knew when the application was delivered.

That was a crazy, exciting day. The day I was finally ready to file. I can't explain how off the charts my emotions were. It was really fun, exciting and scary to send that packet off. I actually was shaking as I handed it to the post office clerk.

There is a step-by-step tutorial for filing a patent via the web on the USPTO.gov site. Navigation for me on the USPTO site is sometimes challenging, but going to the "Filing A Patent on Your Own" page is a helpful place to start gaining more information about the process. As I said before, the purpose of this book is to inspire you. Even if you are going to seek expert help, which is highly advisable, being knowledgeable about the basics of patents is powerful stuff. Just think how awesome it is to walk into a patent attorney's office with a complete draft of a submittal packet. Who knows your invention better than you? If you've built a prototype of your invention, you probably have a great deal of experience and understanding of your invention that a patent attorney most likely may never achieve. You can however, describe your invention in great detail to them and if you have a prototype, you can show it to them. With their expertise, and your honed in knowledge of your invention, great work can be accomplished.

CHAPTER THREE: WHERE DOES CREATIVITY COME FROM?

But, back to the "I'm creative but don't have anything specific on the drawing board" people. If you're really taking this journey. I mean the whole journey; searching patents in a field you like and then exploring how patents are constructed and get filed, then something has to pop right? Is it inventing something that creates inventions? An app or tool that might make creativity flow with ease? Inventions don't always need to be patented and some inventions are better off without a patent as we shall see later in this book.

Even the people that started off on this journey with us that had the billion-dollar idea probably have a few more of them from looking thru patents. They may have made new discoveries that might be just seedlings right now, and they may have found new soil to grow those ideas in and I hope you do too.

Take a look at the top five patents you've done research on. They could be similar to an idea you already have or an invention that piques your interest. What's of importance to them? What useful objective do they serve? Can you go thru and list all the claims in them and find ways to improve upon them?

Let's take a look at patent value. That's another thing to consider. What's the economic value of the patent? For me, I have to try to discern how many people own a jukebox with the particular module my invention is designed to replace and how many of those people are dissatisfied enough to pay for my improvement. Another consideration for me is how much is my cost of production vs. what I can sell the module for? I'm not an economics professor, but I still need to take a swing at the numbers to make sure my invention is marketable and profitable.

It could be a huge company will want your patent and pay you dearly for it, but what if you had to produce and sell it for a while first? Or maybe forever. It could be that a large company won't take the time to understand your invention. You might have to start your own company to produce and market your invention. Many inventors have had that experience. Could you, do it? Produce and market your invention? It's possible that you have a vision that outsiders and investors can't see or won't see.

CHAPTER FOUR:
THE INVENTION NO
ONE WANTED

The story of Xerox is an interesting one and I think it brings some of these issues to the forefront. This is the story of copiers and laser printers and even fax machines. Wait. What's a fax machine? If you don't know, keep reading anyway! Even more interesting is the personal vision quest of the inventor, coming from financial adversity and personal hardship. Both of the inventor's parents became very ill and by the time he was in high school he was the primary financial support for his family and was working for a local printing company. Talk about life problems interfering with creativity, but he overcame it. At the printing company he began to imagine new ways to duplicate printed materials. After a few in between jobs, Chester Floyd Carlson, started working in the patent office of Bell Telephone Labs and then the patent office of P.R. Mallory & Company in New York. In his work, he dreamed of a better way to duplicate the drawings and descriptions required for patent applications, which at the time, were time consuming. He worked at night in an apartment developing what would become the modern-day copier. A machine that could make copies on plain paper. There were other ways of duplicating printed material, but they were costly and didn't always produce a high-quality result. Most of them also required special paper and messy toner.

Chester spent a lot of time at the Public Library reading science and technology books and articles. He came upon one written by a Hungarian physicist by the name of Pal Selenyi.

This was an amazing article that described a way to use an electrostatic charge to transmit ions to a photo conducting powder, just what Chester had envisioned. But the invention of Pal Selenvi needed improvement, Chester decided to use light to remove the charge from areas that had no printing. Chester's invention was called "electrophotography" then, later to be termed "Xerography" and Chester could not get anyone interested in the invention! His process was a dry copy. The technology of that day was a wet copy produced by Photostatic process. He spent years trying to get companies interested in his patent. More than 20 companies passed on his idea. Those companies that passed on such a great invention, included Kodak and IBM. Proof that even experts may not recognize a great invention when they see one.

In spite of the rejections and various industry experts saying it was a "dumb idea", Chester persevered in his endeavor to get funding to produce the copier of his dreams. He was a patent attorney, that studied some chemistry, and the patent claims that he crafted for his new electrophotography invention were so broad that they provided him a great deal of protection from competition. However, his patent was very rough.

Just as Chester was about to give up on producing his invention, an interesting young engineer, Russell Dayton, came into Chester's day job at Mallory to testify as an expert witness in a patent case. Russell was working at the Battelle Memorial Institute in Columbus Ohio.

The two men got to talking and of course, Chester told Russell about his invention.

Russell took Chester's invention back to Battelle and they invited Chester to make a demonstration. Although the demonstration was crude, Russell told his colleges, "This is the first time any of you have seen a reproduction made without any chemical reaction and with a dry process." The Battelle Memorial Institute decided (somewhat reluctantly) to act as an agent for Chester's patents, finance further research and develop the idea. Battelle, like Chester, tried to get major companies interested, but couldn't. Think of that, now with some financial backing and expertise on Chester's side the industry still couldn't see Chester's vision. There were even some engineers at Battelle that felt backing his invention was a mistake.

Then came another miracle. John Dessauer, chief of research for the Haloid Company read an article about Chester's copy process. Haloid was in the duplication business, competing with Kodak, that dominated that industry at the time. Haloid was looking for a way to expand in the duplication field and take market share away from Kodak. In December, 1945, Haloid signed an agreement with Battelle to license Chester's patents for a commercial product. The $10,000 contract that was signed then, is equal to about $ 150,000 in today's money. It represented about 10% of Haloid's annual earnings. A pretty big investment for a small company.

Battelle continued to conduct research under the agreement and the Haloid Company continued to try to produce a commercially viable product from the invention.

Haloid gained a government contract in 1948 from the U.S. Army Signal Corps. This ended up amounting to an investment of $200,000.00, almost 2.5 million in today's dollars to develop a duplication process that would be immune to nuclear radiation. Traditional photographic means used for reconnaissance failed when exposed to radiation.

Over half of Battelle's budget for developing Chester's technology came from government contracts. Without Haloid seeking these contracts, they would not have had the resources to develop Chester's invention.

Battelle at the time had most of the workers devoted to military research. Chester's invention showed promise, but was far from creating a proto-type that worked. They assigned 15 people to work on the invention. His invention used a static charge and depended on a photo conductive material. A type of material that didn't exist. It took many experiments to create an aluminum plate coated with selenium to hold the static charge. Creating the plate itself was a process because it had to be perfectly clean. More than 50 different cleaning agents were tried before finding that plain glass wax that was used in homes to clean windows worked. This was used to clean the aluminum plate before applying the selenium. The static charge also needed to be uniform and at first a sewing needle was used, later to be replaced by fine wires. The fine wires from wire audio recorders were used.

Wire recorders are the predecessor of tape recorders. Instead of using magnetic tape to record sounds, they used thin wires. Thankfully, wire recorders did not last long before being replaced by tape recorders because their sound reproduction quality and dependability was very poor. The invention and evolution of sound recordings is truly interesting as well, but a topic for another time.

Now, came the experiments with the dry toner. This was a completely new product as well. More than 500 tries at it failed. It became difficult to produce a workable product because the toner experiments were affected by temperature and humidity, so depending on the time of year and climate, it may or may not work properly. This had to be considered and solved before rolling out a commercially distributed product. Eventually, of course, a formulation that worked in all climates and most temperatures was created.

In 1949, Haloid shipped the first commercial photocopier. It was called the model A, and was difficult to use.

But, none the less, "Xerography", a term coined by a public relations employee at Battelle, was born. This would eventually open up new markets for products and new printing processes. The new copier would have been a commercial failure, except for its use as making paper masters for offset printing presses used by large corporations. These large corporations, such as Ford Motor Company, kept the process and invention alive by purchasing the innovative invention.

In 1955, Haloid signed an agreement that gave it full title to Chester's patents. Chester had already been working as a consultant at Haloid, but now, the new agreement would give him 20,000 shares of Haloid stock.

Still, in 1957, Haloid could not get any of the large, established corporations interested in Chester's invention. Crazy, right? So, they decided that they would take on production themselves. This is where I was saying, could you take your invention to market? Sell it on Ebay or Amazon? It might come to that even if you have a great idea. In 1958, Haloid changed its name to "Haloid Xerox" and decided they would produce and market Chester's invention. By 1959, the company had created a new photocopier that was economical, easy to use and didn't need special paper. Later, in 1961, Haloid Xerox dropped the Haloid and just became Xerox.

Around 1964, Xerox started marketing the copier that they called the 914. The 914 was fairly expensive to purchase, and they wanted to have a large market share. They came up with the genius idea of leasing the equipment instead of selling it. This would create a dependable financial stream as well as allow more companies to use the invention. By 1965, the 914 was bringing in $ 250 million per year. That's over 2 billion in today's dollars! Deciding to take on production and marketing directly really paid off for them. In fact, Chester's invention now brought in the majority of Xerox's sales. The invention nobody wanted was producing a huge revenue stream for them.

Chester's stock at the time was valued at $500 million, more than 4 billion in today's money. His invention, was a billion-dollar idea. In fact, a multibillion-dollar idea. But that money wasn't his primary motivation. It was his dream of a copier that everyone could afford and use. He never invented anything else, but what an invention he gave the world! I can't imagine businesses working today without a copier, or printer of some sort. I realize we can do a great deal paperless, but I'm not sure scanners would have existed without his dream. I used a copier to create my patent drawings and get them into the right format to submit. Even if I chose a web-based submittal, I needed to create drawings and a copier and laser printer really comes in handy and saves a great deal of time. Because of his job, as a patent attorney, Chester directly realized the need for his invention. Something, big business of the time couldn't see. They couldn't or wouldn't see it even when it was put right under their nose. Some of the companies saw his invention as a threat to their current business. They might lose revenue from their existing copy business, but they were short sighted and couldn't see how a great invention could expand their reproduction business and create new revenue streams.

I love the story of Chester and Xerox and hope it has made you ready to take on the world with a great invention of your own. These stories of great inventors might be too over the top. Most inventors don't have all the resources that Chester eventually had. You can be a great inventor without changing the world. If your invention can just change a little part of the world, then that in itself is a tremendous victory.

CHAPTER FIVE: CLASH
OF THE TITANS

A t the beginning of the book, I mentioned Thomas Edison. For sure, one of America's great inventors, but I find Nikola Tesla much more interesting and inspiring and the strange interactions between Edison and Tesla entertaining and humorous.

Tesla grew up in Europe. His mother was his inspiration. She was extremely gifted and made her own looms and churns and other appliances used in their home in the 1850s. Originally, Tesla was studying to become a professor of mathematics and physics, but the field of engineering soon called to him so strongly, that he couldn't think of anything else.

Most of the electrical systems of that time period worked on direct current or DC. This is the same current that a battery provides. Although AC or alternating current existed and had been developed in principle by Michael Faraday in the early 1800s, most electrical engineers of Tesla's Day had limited experience with it. Motors and generators used a commutator or brushes. An electrical generator was called a dynamo. Tesla's college professors told him motors and generators required commutators or brushes and creating one that didn't was impossible. The other students in the class even made fun of Tesla's idea that a generator could be built without brushes and commutators. Something inside him disagreed greatly with their teachings and he set out to create an electrical generator and motor that worked on a new principal. The thought of this invention was on the forefront of his thought's day and night.

He worked in Hungary in the telegraph and telephone business and later in 1881 moved to Paris and gained employment as an electrical engineer. He kept experimenting with designs for an AC motor and from each failure, and there were many of them, he gained more resolve to succeed. It was in the summer of 1883 that he built his first induction motor. He visualized a device that would have three alternating currents each out of sequence with the other two feeding windings around a stator and creating a rotating magnetic field. He tried to get the Continental Edison Company in Paris interested in his invention, but they used DC power and had no interest in the invention. It was in Paris that he started meeting Americans and having conversations with them about the inventing and development process in the United States. The United States at the time was making news for its remarkable progress in the electrical industry. He learned that the U.S. government had several programs that supported inventors with new ideas. He decided to move to the United States in September of 1884. The journey was not an easy one for him, he had to sell text books and belongings to get enough money for passage to America on a steamship. Before even starting the journey his luggage was stolen, which had his ticket for the steamship in it. Losing your luggage and ticket would have stopped most people dead in their tracks, but not Tesla. Because of his photographic memory, he could "see" the ticket in his mind and recite the number to the steamship crew. When no one else produced a ticket with the same number, he was allowed to board the ship and sail to New York.

Tesla's plan, once in New York, was to arrange a meeting with Thomas Edison.

Although the Continental Edison Company in Paris had no use for his invention, the manager had written a letter of introduction that Tesla planned to give to Edison. Tesla was also going to pitch his invention to Edison. A meeting did take place and Edison said he was not interested in Tesla's invention, but instead, offered him a job. Tesla became an employee of the Edison Works in Manhattan.

Tesla admired Thomas Edison and was very enthusiastic about working for him. Thomas Edison at the time was constructing electrical systems based upon DC current. Early electrical systems were dangerous and started many buildings on fire. Early arc lighting systems were powered by 3000 volts. Edison's dream was to build a safer electrical system. Edison, being a very driven and exacting man, was sometimes difficult to work for. Edison told Tesla that he would pay him $ 50,000 (1.5 million in today's money) if Tesla could design an improved DC generator. When Tesla presented a short core dynamo with double the efficiency of the typical long core dynamos to Edison, he said he was only joking, and did not give Tesla any bonus at all.

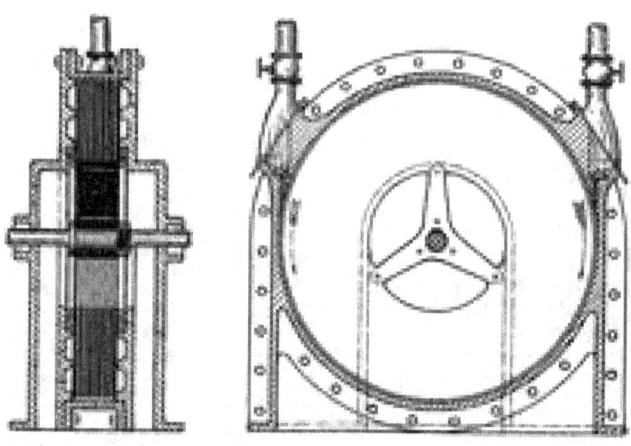

Tesla was very angry and quit shortly afterwards, determined to punish Edison. Tesla's idea, the thing that was on his thoughts day and night was in direct competition to Edison's core business. Even though Tesla had admired Edison, he wasn't going to give up on his dreams and you could bet that Edison wasn't going to change his core business over night, so these two inventors were on a collision course even before they met. Now, with new resolve to punish Edison, Tesla was more driven to developing a safer AC electric grid and show Edison and the world that his generator was indeed superior to those that existed then. In his time working at Edison, he had continued his own experimentation with creating AC generating motors. DC electrical grids like Edison created, are less efficient and limited in how far they can transmit electrical power without a severe voltage drop. Tesla knew in his heart, that he could design a more efficient electrical grid.

Early in 1887, Tesla was offered a chance to start his own company based upon his ideas and inventions and "Tesla Electric Company of New York" was formed. This new company would produce rotary field motors, AC generators and arc lighting. The company ended up failing, but Tesla did not quit on his dreams. In 1887 and 1888 he was granted 35 patents for his inventions. Seven of them in 1888 were for his unique design of the AC generating motor. He addressed the American Institute of Electrical Engineers that same year and George Westinghouse, also an inventor, took notice.

Westinghouse had made a fortune on his invention, an air brake for railroad cars. He, like Tesla, believed in alternating current (AC) as the future for electrical distribution and Westinghouse had built the first AC electrical grid near Boston. Westinghouse and Edison were fierce competitors which made Westinghouse even more attractive to Tesla, who now had an axe to grind with Edison. Tesla also wanted to show the world that his thoughts on the efficiency of AC power were correct.

In 1889, Westinghouse licensed Tesla's patent for the AC generator and hired him. The negotiated royalties that Tesla was receiving on generated electricity were so great, that he then, decided to start his own company. Westinghouse had paid him $ 1,000,000 cash for the patent rights and then an additional $ 1 per horsepower generated. One million then, is about 31 million in today's money's purchasing power and the dollar per horsepower generated is about $ 31 dollars in today's purchasing power. This generous agreement, with a huge payout, became a tremendous burden on Westinghouse, to the point it threatened his company's very survival. Perhaps, Westinghouse was a much better inventor than a businessman. If Westinghouse went under, Tesla wouldn't receive any royalties anyway and Edison would win. Tesla's vision of AC power grids would vanish. Westinghouse explained to him how the original agreement between them was threatening to bankrupt the company and that they needed to renegotiate the agreement to keep Westinghouse from going under. Tesla simply tore up the agreement. It's estimated that Tesla in effect gave up more than $ 372 million in today's money by voiding the agreement. That is, if Westinghouse could have survived financially and made good on what he would have owed Tesla under that agreement. Tesla wanted his invention to be rolled out and used. It was important to him and gratifying to see something he had invented and dreamed about, literally lived for, become a reality that benefited mankind. He also wanted to prove his vision of AC power to all the naysayers and especially to Edison himself. His competition with Edison was so fierce that when in 1912, He was co-named with Edison to win the Nobel Prize, he refused it.

The 1890s were a time of incredible advances and experimentation for Tesla, he invented the Tesla coil.

A high voltage transformer that arcs between things. You've probably seen them in movies a few times or even in person at novelty stores. He also introduced radio communications two years before Guglielmo Marconi. At one point, in Madison Square Gardens, he constructed a water tank in which he floated a radio-controlled boat. He took requests from the audience and used a telegraph key to make the boat respond to requests from the crowd. He also created the first hydro-electric power station at Niagara Falls in cooperation with Westinghouse. Construction started on it in 1893 and was completed in 1895 to become one of the engineering marvels of the world at that time. In 1893 Westinghouse won the bid to electrify the World's Fair and finally AC power grids started to get the upper hand. At the fair, Tesla showed off many of his new electrical gadgets. He sent electricity through his body and astounded the audience by lighting bulbs between his teeth and in his hands. Over the next three years General Electric switched from DC to AC for their electric power grids and AC became the dominant electric power source.

Tesla continued his designs and experiments. He was convinced that truly amazing things happen at higher frequencies of electrical generation, even though his motors operated at lower frequencies. His belief and experimentation with higher frequencies created his discoveries of radio communications and paved the way for all modern communications such as television, cell phones and even radar.

A RADIO PATENT 100th ANNIVERSARY

Microwave ovens and microwave communications exist in part due to his development of high frequencies. He even developed a death ray. In 1895 Tesla's New York Lab burned. This was a huge set back. He lost several years of research notes and experiments, including prototypes. No one, not even Tesla knew what was lost because of this. In 1898, he relocated to Colorado Springs, CO and set up a lab. Colorado Springs had the largest electrical generation plant west of the Mississippi River at the time and was a perfect place to set up his new experiments. He built an 80-foot-high tower with a 200-foot mast and large copper ball on the tip top. He was going to attempt to transmit electric power wirelessly. Light bulbs three hundred miles away with no wiring connections to his lab lit up. Sparks flew from water faucets all over Colorado Springs and then stopped. The bulbs went out too. His experiment had fried the electrical generator at the Colorado Springs Electric Company. In 1900, he went back east to Long Island and with financial backing from a group of investors, constructed a lab and built a huge transmission tower in the village of Shoreham in New York. The tower was called Wardenclyffe.

A group of investors had kicked in $300,000 (about $ 10 million in today's money) to back his idea of transmitting free power to the masses. After many delays and the obvious economic issue of how to make a profit on free energy, the investors pulled out of the project and the lab wasn't completed.

So, some of the things to think about is Tesla did not invent AC power, he didn't even invent the first AC generator or motor, he created an engineering marvel at the time by creating a unique AC motor and generator that made AC power practical and more affordable. Most inventions are refinements of earlier ones. They are not completely new technology or thoughts. Tesla also patented a transformer. Transformers are constructed of wires that are wound around a metal core, they can jump up the output voltage as in the case of a step-up transformer or reduce the output voltage as in the case of a step-down transformer. At higher voltages, wiring can actually be smaller and there is less voltage loss making distribution of electricity more efficient over long distances. Once the voltage is nearer to the destination it can be reduced to household levels by use of a transformer. This was one of the early advances in electrical distribution and the design and construction of electrical grids.

CHAPTER SIX: WHAT IS REAL ART?

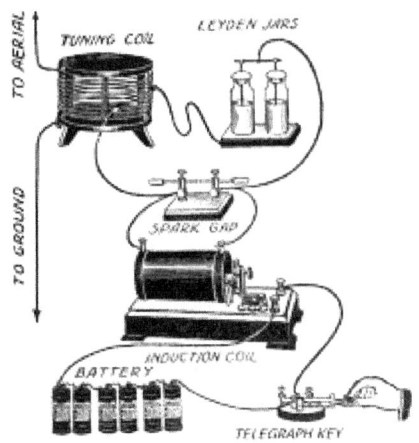

W hen speaking about patents and art (the drawings or illustrations of patents), we talk about improvements over prior art most often. This is because someone's generally created something already in that field and the new invention improves upon it. It's how society and technology evolve. And it is vital to have creativity and improvement and to foster it. So, if you have a great idea and start exploring patents and see others thinking some of your thoughts, great! It means you share creative genius with others that have already secured a patent. Does it mean that your idea is dead? Should you stop your efforts? Not always! Focus on the improvements your art has over previous art. What makes your invention stand out? This is going to be the basis for your claims that will be listed in your patent application.

If you decide to file your own patent application, it is extremely helpful to have your own precise and clear thoughts of what already exists from other inventors and how your patent improves upon prior art. The same can be true if you use a patent attorney, why not bring them a well thought out invention? You can have a multitude of advisors and opinions which could be helpful, but the bottom line is no one knows your invention like you do.

Let's go back again to the construction of a Provisional Patent. Remember, this is a patent for just 12 months and will be abandoned if you don't file a regular or Non-Provisional Patent before the 12 months expire.

In writing my Specification, there were several elements;

1)   **Field Of Invention**, which describes what the invention relates to. For me this was jukeboxes and more specifically jukebox interfaces.

2)   **Background of Invention**. I described how I found a need for the invention, things I noticed about existing devices and their deficiencies and what I did to overcome the deficiencies.

3)   **Summary Of the Invention**. I described what the invention was comprised of and then described each of the drawings and what they depicted.

4)   **Relationship Between the Parts of The Invention**. I described various interactions between parts of the drawings.

5)   **Details Of Operation and Functions**. Here I made a brief summary of how all of the parts work together.

6)   **Claims**.   This   was   on   its   own   page   as   required   in   the   instructions   on   the   USPTO

site. I found some of the instructions at: https://www.uspto.gov/patents/basics/types-patent-applications/provisional-application-patent and tried to follow them as closely as I knew how. If you decide to file an application on your own, which I don't advise, you will want to read as much information on the subject as possible and become as informed as possible.

7) **Abstract Of Disclosure**. This was my summary of the invention. I wrote what the invention does and how it is implemented. The abstract is limited to 150 words.

The reason why I want to go back to the Specification is to look at it much closer now that you have thought about patents more. Going thru the Specification and writing about your invention is not a waste of time. It is a fact-finding mission that really should be quite fun. Many inventors live in their heads, which is great for the design phase, but at some point, they will need to translate thoughts into action and into a physical format that allows them to gain a patent (if they want their invention patented). Taking into consideration the Specification and actually sitting down and putting something down on paper or on a computer or tablet or phone can really help. It made me focus more and created a structure for the development of prototypes and then aided in the actual patent application process. Having a draft of the Specification can also provide your patent attorney or patent service vital information. For me, it also created new and original thoughts and provided extra design improvements to my art and prototypes.

In going to the patent office site: https://www.uspto.gov/patents/basics/types-patent-applications/nonprovisional-utility-patent#heading-9 at the uspto.gov and reading the Non-Provisional Patent Filing Guide gave me a real break down of the Specification. This was helpful in filing a Provisional Patent because the next step for most inventors and more specifically me, is to submit a Regular or Non-Provisional Patent Application claiming benefit from the previous Provisional Patent. If I want to claim that benefit, there can be no new claims included in the Non-Provisional Application. So, looking at the Non-Provisional instructions was very helpful in attempting to have a good foundation from the Provisional Application.

CHAPTER SEVEN: THE INVENTOR'S ISLAND

M any inventors refuse to speak to others about their invention until they have a patent and I personally want to avoid talking to people, even family and friends about specifics of my inventions. You may feel the same and feel like you are on an island without support, but you can get much needed and sometimes free support at: uspto.gov and searching "Learning and Resources, Inventors and Entrepreneurs." One of the pages I used and of course the links are subject to change, was (and is): https://www.uspto.gov/learning-and-resources/inventors-entrepreneurs-resources (If the links change just go to the main web site: www.uspto.gov and use the search box for the topic you want). If you are careful about what you say or disclose, you can talk to other inventors on social media and family and friends in general terms about your ideas without jeopardizing your core invention. But, consult with your legal advisor before doing so. Family, friends and others may be helpful, but don't always take to heart everything they say either, they may not be able to see the benefits of your invention as you see them.

I filed everything by paper thru the US mail for my invention. I really felt some connection with inventors of the past by using that process to file. It's crazy to think that the old inventors that we have discussed drew all their drawings by hand without computer aided anything. Some of them typed out Specifications on a type writer. Some didn't even have that. Right now, you can submit a Provisional Application on paper thru the mail without additional filing costs. This is not true for filing a Non-Provisional Application however. The Patent Office is trying to discourage filing by mail. If you choose to file a patent application by mail, instead of electronically filing, you will encounter additional filing fees that are fairly punitive.

So, the point is these inventors had to have the tenacity to overcome huge obstacles as we have discussed and then work much harder at perfecting a patent application than we do. This makes me want to try even harder when something or someone tries to stop me from continuing the day-to-day struggle with my own obstacles. If my predecessors could create inventions without a computer and the technological gains we have now, then why should I be defeated so easily? Hopefully, you will not let anything stop you from succeeding in your goals. If they are to get a patent on your amazing and unique idea, then forge onward. If they are to start manufacturing or marketing of your protype, then by all means don't let this chaotic world stand in your way.

CHAPTER EIGHT: PRANKSTERS THAT BECOME INVENTORS

Maybe the inventors that I have talked about don't inspire you. Or perhaps they do, but "that was then, this is now," right? Anyone we discuss will be somewhat past tense, but let's try somewhat more current inventors.

Are you familiar with Stephen Gary Wozniak? That name doesn't immediately equate to most people as a famous inventor, let alone a person that changed the face of technology in a major way. But he is an inventor that did just that. He was expelled from the University Of Colorado Boulder for hacking the universities' computer system.

He then enrolled in De Anza College and later transferred to the University of California, Berkeley. In June of 1971 with his friend Bill Fernandez, they built a computer that used punch cards as a self-taught engineering project. This device was very limited in comparison to what a person expects from a computer now. It was comprised of 20 TTL chips that were given to them by a friend. At an exhibit, a newspaper reporter stepped on the power supply cable and shorted out the computer and it was never functional again, but the ground had been laid for further computer designs.

Bill Fernandez had a friend that "liked electronics and liked to play pranks" and thought Wozniak and him would get along well, so he introduced Wozniak to Steve Jobs. Steve Jobs was still in high school at the time, that was way back in 1971.

Wozniak dropped out of Berkeley and gained employment at Hewlett Packard designing calculators, where coincidentally Steve Jobs had a summer job working with main frame computers. So, Wozniak is a drop out that also was earlier expelled from another college. Interesting right? Does it remind you of another inventor that was told they wouldn't amount to anything? Or, perhaps, a string of successful people that refused to quit just because some people refused to see their value.

In October of 1971, Wozniak read an article in the Enquirer called "The Secrets of The Little Blue Box" by Ron Rosenbaum which allowed someone to make long distance calls for free. The technology is called "Phreaking", which is hacking into a telephone network to avoid paying for long distance calls. This was a time when people called each other and there wasn't VOIP (Voice Over Internet Protocol) services, so calling family and friends far away meant paying the phone company by the minute and it was expensive.

The inventor of the "Blue Box" is anonymous. They never got a patent on the technology because of the questionable legality of it. For sure, stealing long distance phone service from the phone company is illegal, but manufacturing the "Blue Box", I'm not so sure. If it's sole use and intent is a device to steal phone service it most likely is illegal. If the device can be used for other things? Is it a grey area? A question for attorneys and I'm not one.

But the "Blue Box" story from an inventor's prospective is super interesting and I'm going to make a segway here and take you on that journey. Before you can take it though, you have to somehow transport your mind back to 1971. People have to pay the phone company for monthly service and then additional charges for long distance. The phone company consists of a monopoly, there are no alternative communications for calling or communicating with people long distance except maybe HAM radios. Amateur Radio (HAM radio) was and is a popular hobby and service where radio operators get an FCC license and communicate with each other.   It is its own social network now and it was so back in the 1970s.   This social network is a pretty small segment of the population. The internet, cell phones, personal computers and social media don't exist. So, most social interaction happens face to face or by telephone.   Land Line telephones. How many times a day do you look at your cell phone? Not in 1971! There exists a love-hate relationship that most people have for the phone company. They love to talk to other people and use the technology for business dealings, but hate paying their bills and when the service is down or malfunctions, they are very frustrated because they have no alternative. You can't quit the phone company and choose another provider, you are stuck with them!

The first thing that I find interesting is most of the "Blue Boxes" made were not really blue.

No one has given an undisputable reason why the name was coined for the device, not even the inventor. It could be that some of the first ones that were built, were built using the blue Radio Shack project boxes of the time, but there isn't a definite account of where the term came from. Most of them that were produced over time were actually black, grey or blackish grey. Some were made out of electronic project boxes and other electronic parts that were purchased from vendors such as Radio Shack and Lafayette Radio Electronics.

People using the "Blue Box" technology were called "Phreakers" and what they were doing was called "Phreaking". So, the Phreaking Phreakers (Sorry, I find that hilarious) used a technology developed by an anonymous inventor, we'll call Einstein for our purposes. Einstein was in many respects incredibly talented and definitely an inventor. He read, in a technical journal, an article that a Bell Telephone engineer published about switching theory (connecting calls) that also listed all of the telephone companies' multi-frequency tones. Allowing the public release of this information is akin to publishing the formula for Coke. The telephone company tried to remove all copies of the technical journal, but it was too late.

Almost immediately after reading it, Einstein started engineering a prototype "Blue Box" and had a working device within twelve hours.

The device he created was basically a glorified tone generator. Not a very sophisticated device, even for the 1970s. Tone generators existed in many forms.

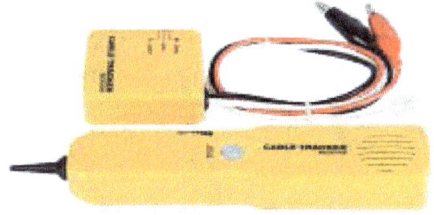

They were and are used by telephone company repair men to trace lines. Tone generators, as the name implies, create audio tones. The trick, or the genius was to create a device that generated the correct multi-frequency tones from the technical journal. Devices communicating over phone lines create audio tones. Think fax machines, computer modems, security systems, push button phones etc. The same is true for the telephone companies' switching technology itself, it is dependent upon tones. This is the technology that decides where and how to route phone calls. The telephone company at some point in the early 1950s made a decision to operate their entire switching system on twelve electronically generated combinations of six master tones. The multi-frequency tones are fairly simple. For instance, 900 cycles and 700 cycles at the same time produces a "1" at the telephone companies' switch.

Bell Telephone's toll network was and still to some extent is, comprised of hundreds of toll switching offices connected with trunk lines. Each toll switching office has thousands of long-distance tandems.

A tandem is a switch (think of an ordinary light switch connecting a light bulb to electric power) used to connect trunk lines and also interconnect to other tandem switches which in turn connect to more trunk lines. Trunk lines are bundles of wires running between destinations. When dialing a long-distance number, a tandem listens to your request and connects you into the telephone network and thru other tandems and trunk lines if needed, and routes your call to the proper country, state, city and then the person's or companies' phone number you are calling. The tandems "speak" to each other and route calls. The speaking part, means that they can hear audio tones between themselves.

Einstein, understanding how the tandems worked, knew that when the tandems were idle, not in use with a telephone call, they generated a 2600 cycle per second tone. He discovered the tone's precise frequency from the technical journal. This frequency lets other tandems know that it is not in use and can be utilized for making long distance calls. The other usable tandems in turn throughout the network are also generating this 2600 cycle tone. Once a long-distance number is dialed, the tandem that you are connected to stops sounding the 2600 cycle tone so that the rest of the network knows it is no longer available for use by another caller.

To hack, or Phreak the technology, a Phreaker calls a long distance number. To keep from getting charged, they would call a 1-800 (toll free) number on their home or pay phone. Using the "Blue Box" the Phreaker then held up the speaker that was on the "Blue Box" itself (many of the boxes just had the ear piece element from a telephone for a speaker) to the mouthpiece (microphone) on their telephone and pushed a button that generated the 2600 cycle tone. This immediately interrupted the 1-800 call that they made. The tandem now thought that the call was completed and that it was idle, it was now waiting for a new call. When the Phreak then released the 2600 cycle button on the "Blue Box" the tandem then showed in use and waited for instructions via tone as to where to route a call. The Phreaker just dialed a new long distance number using the keypad on the "Blue Box" and the call was connected. To the phone company the Phreaker called a 1-800 number and then hung up when the Phreaked call was completed.

Once Einstein started selling "Blue Boxes" he found there was a huge market for them. So much so, that Einstein was selling thousands of them and having them made in the Philippines. Many others started making the devices. But more than just a market for the devices, up sprung a sub-culture of Phreakers. To some, they were battling for the injustices of a huge tyrant called Bell Telephone, to others they just liked the social network that was being created between them. Phreakers would talk between themselves and conference calls between cities and countries. If you couldn't afford a "Blue Box", which could cost as much as $ 1500.00 ($ 11,037.00 in today's money!), you could use other means to generate the tones. Some Phreakers simply used a cassette recorder with a tape that had all of the frequencies recorded on it. You could then create a tape recording of the phone number you wanted to dial from there and simply play it. The cassette tapes with the frequencies on them were marketed quite heavily. There were even instructions on using a home organ or keyboard to generate the proper tones. One Phreaker found that a whistle that was given away as a toy in boxes of Captain Crunch cereal generated a perfect 2600 cycle per second tone. He became known as "Captain Crunch" in Phreaker society.

Back to the law, Phreakers for sure were breaking the law and although it seemed like a victimless crime, it was costing the phone company thousands, if not millions in lost revenue. No one can actually tally the exact amounts they lost in revenue. The phone company created task forces to hunt down and prosecute Phreakers and many were arrested. The task forces created software that looked for patterns in phone use analogous to Phreakers phone use patterns. How many people talk to Avis on Avis's 800 line for two hours? If the phone company had the software set up at the Phreakers' home switching office and the

Phreaker made the mistake of making their calls from home and not a public pay phone, then they ran the risk of being targeted as a possible Phreaker and arrested by the police. Many arrests did occur.

It's hard to say how much Einstein made off of his invention. The Rosenbaum article said that he had a single order for $300,000.00 (2.2 million dollars today). That was a single order and Einstein had many other orders. He had set up off shore manufacturing to keep up with orders. The total sales generated had to make it a worth while venture even though he couldn't patent his invention. And of course, because he couldn't patent his invention there were all kinds of competitors making the devices.

CHAPTER NINE:
BILLON DOLLAR IDEAS

T he article got Wozniak thinking, "I can build my own blue boxes" and so he designed and built some. In fact, he built over two hundred of them and Steve Jobs marketed them. Designing and making these devices was not a difficult task for Wozniak, who by all accounts is a technological genius. They sold them for $150.00 each. A far cry from what the devices started out selling for, but as with any technology once competition and supply grows the price decreases. This created the partnership for further business ventures. Wozniak has said without the "Blue Boxes", Apple may not have happened.

In 1973, while working for Atari, Jobs asked Wozniak to help him engineer a new motherboard for the computer game "Breakout."

The company offered an incentive to make the new motherboard less expensive to produce. The fewer electronic chips used in the design, the better, so for every chip that could be eliminated, they would pay $ 100. That's about $ 673.00 per chip in today's money. The new board design by Wozniak had 50 fewer chips. Atari paid Jobs a $ 5,000 bonus but Jobs told Wozniak he only received $ 700.00 which he split with him. It took Wozniak ten years before he learned about the lie.

As a member of the Homebrew Computer Club in 1975, Wozniak developed a computer to impress the other club members. Most of the computer clubs centered around the first commercially produced personal computer, the Altair 8800.

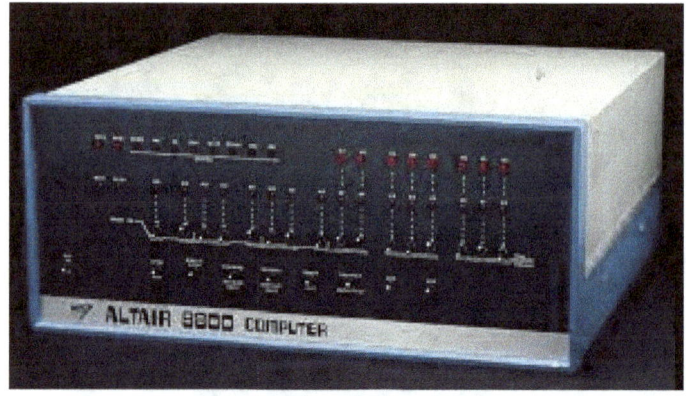

The computer Wozniak developed was unique because of its user interface. Every time it was shown to people it would draw a crowd. Steve Jobs saw it and thought it could be marketable. It was from this device, this prototype, that Wozniak refined and created the Apple I.

Neither Jobs or Wozniak were convinced that this device could be sold profitably but they decided to join forces once again and form a company to sell them. Of course, Apple Computer Company was then created. The name came from Jobs spending time in Oregon on an apple orchard. In order to get enough capital to start, Wozniak sold his HP scientific calculator (they used to be very expensive) and Jobs sold his Volkswagen van. They decided if they failed, at least they'd have a great story to tell their grandchildren.

Together, Jobs and Wozniak assembled 50 of the Apple I computers and sold them for $ 666.66 to Paul Terrell who owned a computer shop in Mountain View, California. He had seen a prototype demonstrated at the computer club where he also was a member. Wozniak came up with the price, because he liked repeating digits. Some might find the number as evil, but it was the start of an outstanding company!

The Apple I was similar to another computer being sold at the time, the Altair 8800. The Apple I lacked a case, power supply, keyboard and display. The user had to provide all of these things. So, Apple Is vary greatly in their appearance. A total of about 200 of the Apple Is were produced and an example of one is at the Smithsonian Museum. While the Apple I was a mild commercial success, it demonstrated that a personal computer could be commercially viable and the new company attracted an investor. The investor, Mike Markkula was an electrical engineer that had business experience as well, which helped form the initial management structure at Apple. His personal investment of $ 91,000 and a $ 250,000 line of credit that he arranged gained him 26% of the company. Jobs and Wozniak each had shares equal to 26% of the company as well. Markkula insisted that Wozniak quit Hewlett Packard and work full time at Apple. He also formed a corporate structure and brought in Michael Scott as president and the first CEO.

The new funding and management structure allowed Wozniak in his full-time role to be even more creative. He thought of his time working on the Atari motherboard and how great it would be to have a computer that could display color graphics. Wozniak began working on the design for a new computer. He used a chip to put color into the NTSC system. It produced colors where a dot appears on a line. To this day he doesn't understand how it works, it just works! It is kind of a side effect. Maybe an unintended consequence that for once does something beneficial instead of the normal detrimental outcome.

The new computer also needed an easy computer language that novice computer users could access and write programs for. The BASIC programming language was the choice of the time and it was going to be built into the new device. There also would be expansion slots for extra memory and adding other functionality and features. This was the birth of the Apple II; a true technological game changer and it was introduced to the world in April 1977 at the first West Coast Computer Faire in San Francisco. The annual event became the biggest computer show in the world, until its demise in 1991.

The Apple II became one of the first successfully mass-produced computers. The first mass produced personal computer, the Altaire 8800 had a production level of about 25,000. While, the Apple II was produced for sixteen years and achieved a production level of more than 6 million units. In 1983 alone, Apple manufactured and sold one million of the Apple II computers. Truly this device helped create and shape the personal computer market.

But, Wozniak originally envisioned that the Apple II would be purchased by electronic hobbyists and others for home use. His invention actually was primarily purchased by small businesses which accounted for more than 90% of its sales.

Wozniak also designed the Disk II floppy drive for the Apple II which was introduced in 1978.

The Disk II replaced the much slower audio cassette recorders of the time as a faster and more dependable data storage device. Early computers utilized cassette recorders to store data after the computer was turned off. The cassette tape cartridges could be inserted and used to store such things as inventory databases, documents, etc.

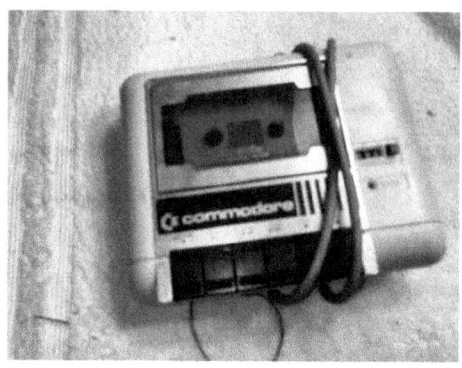

This method of storage was economical but retrieving data was very slow and subject to all the issues a person had with cassette tapes, including the cassette malfunctioning and the tape twisting, breaking, winding up on itself and the many other frailties the technology presented. Of course, an inventors' art and specification could demonstrate all of the improvements the floppy disk drive had over the cassette tape storage system. Think about them.

The floppy disk drive allows the user to seek information faster. Cassette tape malfunction is eliminated. The user's data is safer. Wozniak's design was hailed as an engineering marvel. Not just because it was one of the early floppy drives, but because of his excellent electronic design. He used fewer electronic parts than other similar devices and created a device that was a dependable storage system. This was one of Wozniak's engineering hallmarks, efficient component use and design. Think of just those improvements over existing art? Less costly to manufacture. Fewer parts generally mean lower component failure rates and greater reliability.

The new floppy drive paved the way for the introduction of Apple Dos in 1978. Wozniak didn't write the code for it but the new software needed his device. DOS stands for "Disk Operating System". The drive needed it to function, but arguably, both the code and drive were needed to make the Apple II so successful and allow software developers the ability to write the more than

1500 programs that were eventually released for the Apple II. Think about all the creativity amassed by the various software developers that contributed into making the computer a success as well.

The Apple III was introduced in 1980 and the company believed that it would kill the sales of the Apple II within 6 months. The Apple III was a device designed by committee and a host of engineers instead of Wozniak.

Because of the marketing demographics of the Apple II, Apple believed designing a computer like the Apple III, that targeted the small business segment was essential, so they put a great deal of resources into developing it. But, guess what? All that brain power and resources created a market failure. The failure started with the design itself. Overheating problems and component failures lead to a recall of the first 14,000 units sold and a huge embarrassment for the company. Over confidence and marketing issues compounded the misstep. Eventually discontinued April 24, 1984, the Apple III achieved only about 120,000 total units sold. And of course, the Apple II sales continued to comprise 85% of Apple revenue during those years.

In 1984 the first Mac was introduced.

Wozniak was also instrumental in developing the first Apple Mac. He didn't conceive the design; in fact, it was a design that Jef Raskin conceived. Raskin was an employee of Apple and primarily wrote technical manuals. He wanted a powerful, user-friendly device that was low cost. He came up with the name from his favorite apple. But he was not an electronic engineer and that is where Wozniak came in. Wozniak's engineering aptitude and genius was an essential ingredient in making the new device successful.

When the first Mac was introduced in 1984, the company started focusing most of its advertising and attention on it. Wozniak left the company a year later and sales of the Mac were lagging for many years, the Apple II continued to provide 85% of Apple sales until it was discontinued October 15, 1993.

Apple couldn't will the success of the Mac, any more than they could will the success of the Apple III. It was only thru continuing design improvements that the Mac eventually became a success. Perhaps it was a product release before its time. In market terms the early Mac was a huge failure, but Apple persisted with it in spite of that. Wozniak's early design can be said to have paved the way, but he and Steve Jobs were no longer at the company when other visionaries' marketing and design strategies made the Mac a success story.

CHAPTER TEN:
WHO GETS YOU?

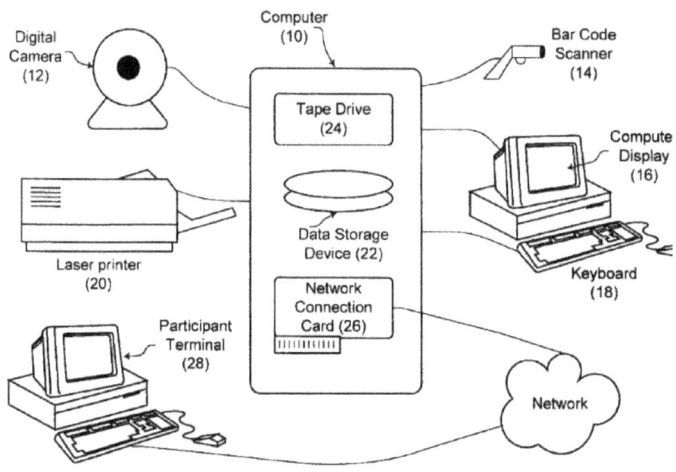

It would be nice if we could all write out our own map for success and the universe would comply. In the stories I've told you of inventor's journeys, the inventor's efforts have been riddled with set backs and surprise twists. It's only thru the strong determination of each of them that their invention came to reality and to market. Most of them needed some good luck and the help of others, and some of them could not have succeeded without a whole crew of support staff.

Chester's Xerox copier was an amazing idea, but remember there was someone else with a similar idea. His invention, his unique perspective of solving a problem, was still very rough and not marketable until someone else saw what he saw and gave him the resources to make it marketable. And even with greater resources, it took years for the market to see what they saw. Chester had the unique perspective of dealing with patent art and patent applications and wanted a way to make his job and that of millions of other people easier. You can see from his story that not everyone he explained it to got it. In fact, very few of the people he explained it to would get it. That's why you as an inventor, have the unique perspective that no patent attorney or patent firm can have. If you find a really great firm or attorney to help you, that is an invaluable resource, but you still can convey your dream, your invention, your art, better than anyone. And that is the purpose of this book. I can't show you or tell you what you should do to protect your vision. I don't have the qualifications to do that. I'm just another person like you, trying to navigate all of that. What I do want to propose to you are different ideas that may help you in that endeavor. By walking in the steps of some of the greatest inventors of our time, you may get charged up and ready for all of the disappointments and successes that you may face. By looking at the requirements for filing patents you may have some ideas of what you are looking for in a patent attorney or firm and hopefully you may avoid some of the scams that are out there.

It wouldn't hurt your cause either to be able to put your patent into a rough draft and start thinking of your idea as art. Putting your invention down on paper and describing it in writing before you try to engage professional help can make it easier to present and could save you some expense. It can also help document your development of your invention and create a timeline that may be useful if you have to defend your invention later. Searching prior art can also save you time and money by avoiding filing a patent for something that has already been done. Prior art searches provide an enormous amount of information if you approach them in a learning spirit.

If conducted correctly, a prior art search will exhibit the necessary elements to similar patents. By examining the patents contained in the search you can see how a patent is properly constructed. Going thru the patent claims allows you to compare the claims of a particular patent to your art and allows you to think in a broader way. It may even lead to improvements to your design that you have never thought of. It may lead to the evolution of your design.

An invention though, is much more than just art. Most inventions are conceived to benefit others and most inventors have a profit motive as well. Chester wanted to help others, but his family was living pretty much up to his income. They had little savings and they were living you might say "paycheck to paycheck." So, he wanted to make money from his invention as well as help people. He had both a benevolent agenda and a somewhat selfish one. Although providing for one's family isn't really selfish. But, the economics side of that equation, making money from one's invention usually requires marketing skills. It's quite unique when someone possesses all of the skills necessary for success in business. Being able to create a business platform, or invention and marketing it as well.

That's why we see the partnership of Wozniak and Jobs being so successful. And Chester needed Haloid and their marketing and organizational skills. Chester also needed customers and those were hard to find at first. It really took the U.S. Government getting interested in his invention to provide the majority of the funding to perfect his invention. And it took the people at Haloid to recognize the potential of his invention's uses for the government to get the funding from them. In many respects, this demonstrates the skill and I'm going to say luck that was needed to bring a truly revolutionary and highly important invention to market and to success.

Just think about all the struggles and the millions of dollars it took for the market to realize how wonderful Chester's invention was and is.

C. F. CARLSON   2,297,691

ELECTROPHOTOGRAPHY

Filed April 4, 1939

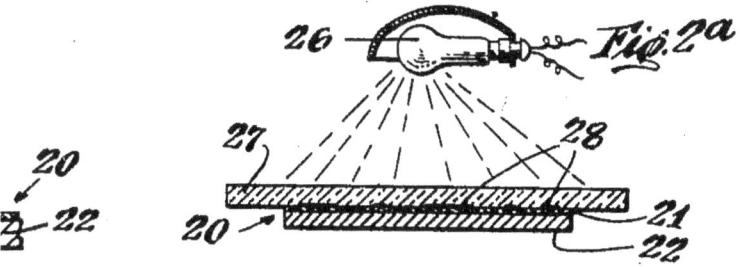

We take his invention for granted today, just as we take a light bulb for granted. But, if Haloid and Chester hadn't kept fighting the good fight and if they didn't also have some good luck in the process, the world might be a totally different place. His invention has shaped modern publishing and business operations of all sorts today, which might be conducted differently if his invention hadn't eventually succeeded. In many ways, layers of new inventions we take for granted today, are built upon his invention. So, even with a great idea, we can see an inventor may struggle mightily to create a marketing success.

Does that scare you? Intimidate you? Depress You? Don't let it. Let it inspire you and just know that you will have to come correct when fighting your good fight. You probably won't get from point a to point b in a straight line. Point A being the invention, the idea itself and point B being a patented, successful invention. Hopefully, you prepare for a long struggle and your ingenuity, skill and luck pave the way for a much easier path than you think. But, do not be dismayed if that doesn't happen. Only you can decide if the struggle is worth it. And that's why I believe directly doing patent searches and research by the inventor can benefit the inventor in ways they may not see before undertaking such a journey. It's a common thing for me that I think "that won't help doing that", or "what bother with that?" and I trudge thru anyway and I discover something that I never expected. What I think "is not worth the effort" turns out to be greatly beneficial in ways I could never predict.

CHAPTER ELEVEN: WRAPPING UP

The inventor I called Einstein, created a product called the "Blue Box" which was used for criminal activity and probably was not patentable. But he is no less an inventor. Just because you don't patent an idea, doesn't make it any less of an invention. And he sold several thousand devices. Let's pretend for a moment, that he spent the 22 months or so perfecting a patent on his invention, he might have missed the market. Without the phone company charging huge amounts for long distance and being a monopoly, I'm not sure his device would have been the marketing success it was. Today, we have social networks galore and a multitude of communication devices, but at that time, there was a void in avenues for social networking, a void that his device filled. Phreakers associated and created an early social network. Other people also noticed the publication that exposed the phone company frequencies so the clock was ticking on how long it would take others to invent competing devices. So, being first to market without a patent proved successful for Einstein. In spite of incredible competition that the inventor actually created for himself, being first to market and his innovation in manufacturing, marketing and design proved to give him a leg up. This is where expert marketing and patent advise is needed by everyone. Finding that advice and knowing how to use it is one of the challenges that you will face and need to solve.

And how about Tesla? He made millions, amounting to billions in today's money but died in near poverty. He spent all of his earnings on his ideas. His inventions serve all of us, even today and we take his sacrifice and ingenuity for granted as well.

Is Tesla's story one of a brilliant inventor that was foolish with money? Perhaps. But without some of his research and development and the use of those funds he wouldn't have progressed some of his ideas. I think his story can be one of "a fool and his money are soon parted" and "it takes money to make money." Many things do not have a rigid dichotomy and do not translate well into two columns "yes", or "no." Could Tesla have accomplished everything he did by not spending his money on progressing his inventions? No, he needed to be the person he was to be able to create everything he did. But, could he have benefited from having someone that understood finances better, understood marketing better. Yes, most certainly he could have.

Often times, Tesla is made a hero in history while Edison is depicted as somewhat of a villain. I won't get into that too much because each person had a certain stubbornness and hard headed look at things that made them the determined and successful individuals that they were. But Edison's methods of invention included creating a business structure to fund his endeavors and a staff to help. Many of the inventions that Edison made were not solely of his efforts, but that of many. Edison did not die in poverty, but rather had a significant net worth. Edison's way of organization created a structure that not only benefited him, but millions of others that benefit from the inventions he was able to develop because he had the funds to stay in business and create them.

I don't believe that we can just look at an invention and say that is great idea, without also contemplating, "how can we market this?" And that is not a selfish question. It would be nice if everyone could magically have all the money, possessions, enjoyment and peace of mind that they desired, but we know reality comes crashing in on that daydream. Somehow, someway there is an economic reality to everything. If you are doing great humanitarian work, you still need funding. It might be raising money for a good cause, but you still need fundraisers to get the funding. And even if your invention can save the planet, it needs to be marketed and to get someone interested that can produce it, or fund the production of it.

Most inventors are loath to speak about one of their inventions before they have perfected a patent. This can make it difficult to create marketing forecasts and judge the economic value of it. Wozniak and Jobs had no idea if their first Apple computer would sell. And I think most people would tag Jobs as being a marketing genius. But they put the product out there and sold around 200 of them. That early venture combined with the sales of the "Blue Boxes" was a full on business adventure for them for sure, but it also was a successful marketing and product production template for them. It showed the partnership worked as a business venture and that Wozniak's inventions were desirable by the public. Job's early marketing skills were utilized and refined. This created the perfect system that made the Apple II such a success and funded the creation of a huge company. A company that I will argue benefits millions of people daily.

A Provisional Patent can be the solution to having something in place that protects the inventor while they seek advice, funding and other needed resources to bring their invention into reality and into the market successfully. It is one of the reasons this application exists and can provide you with some comfort moving forward with business plans. This patent will only provide intellectual protection for 12 months and it will need to be perfected in a way that a Non-Provisional or regular patent can assume benefit from it. And, as I've harped on, you should seek a qualified source to assist or completely file the Provisional Application to assure it covers everything you will need to protect you and gain benefit when you file the Non-Provisional Patent. No new claims can be made in the Non-Provisional Application if the Non-Provisional application claims benefit of a previously filed Provisional Application. You can easily file and receive a

Provisional Patent, but remember the trick is having a Provisional Patent that is beneficial to you later.

So, how can you feel confident in disclosing your invention to someone to secure help filing the Provisional Application in the first place? First off, by reading this book, you have at least a rough idea of the process. Next, by reading all you can on the uspto.gov site and other reputable sources you can gain more information. It is also beneficial to utilize the inventor's assistance center on the uspto.gov site, which you can look at: https://www.uspto.gov/learning-and-resources/support-centers/inventors-assistance-center-iac

You are not alone, because the assistance center has many inventors that have been through some of the steps you will be taking and can help. Be careful when searching for patent assistance and make sure you stick to the U.S. patent office and other reputable sources on the internet. Just searching for patent help will bring up many sites which may lead you into getting involved in a scam.

Once you have perfected a Provisional Patent that will stand up to legal challenges and be able to claim benefit without question by a Non-Provisional Patent, you will have a date that your patent protection starts at. This would be the filing date, the date you actually mailed or filed the Provisional Application on the internet with the U.S. patent office. Having a U.S. patent however, does not protect you outside the U.S. and you will want to explore filing requirements of other countries to gain those protections. You can find more information on that by searching other patent offices throughout the world and by talking with your patent attorney or patent firm if you have engaged one.

If you have satisfied your fears of discussing your idea freely, you will probably want to try some marketing tests and evaluate your idea in terms of economics and see how you will proceed to bring the invention to market. You may also need to consult with a firm or person that have unique marketing qualifications for this. You will also want to be discriminating in your selection of a marking firm or individual because many scams exist in that realm as well.

The burden of protecting your intellectual property and creating a way to bring it to market firmly rests with you. Remember that, you need to be an active participant and can even enjoy and should enjoy the whole journey. Wozniak left Apple after he had achieved immense success because he wanted to go back to the early days of invention. Back to the fun of experimentation, research and development of ideas and inventing. The journey was a lot of fun for him and can be for you as well, so enjoy it. You may find that your idea becomes something that fosters and creates an industry, or a large business model. You might find yourself looking out the window and reminiscing over the steps you took in getting there. So, enjoy the small successes along the way and don't let the struggles in getting to the finish line defeat you. A great success is what I wish for you, but more than that I also hope that you enjoy every minute of the pursuit of it and the process in achieving it.

www.ingramcontent.com/pod-product-compliance
Lightning Source LLC
Chambersburg PA
CBHW071026220526
45467CB00004B/1528